Bibliographic information published by the German National Library:

The German National Library lists this publication in the National Bibliography; detailed bibliographic data are available on the Internet at http://dnb.dnb.de .

Imprint:

Print and binding: Books on Demand GmbH, Norderstedt Germany
ISBN: 9783668927421

This book at GRIN:

https://www.grin.com/document/456043

Ishita Bharadia

Effectiveness of oxidizing reagents in removal of blood stains

GRIN Verlag

To what extent does the change in mass (1- 5g) of the oxidizing reagent, sodium per borate ($NaBO_3 \cdot H_2O$), sodium per carbonate ($Na_2 CO_3 \cdot 1.5H_2O_2$) and hydrogen peroxide (H_2O_2) found in detergents is effective in removing bloodstains, measured by spectro-photometric analysis of change in chemiluminescence intensity of luminol due to the oxidation of ferrous (Fe^{2+}) to ferric ions (Fe^{3+}) in stock solution containing oxidizing agents and iron solution (120 µg/dL) imitating blood at room temperature (30°C)?

Table of Contents

1. Topic

Effectiveness of oxidizing reagents in removal of blood stains

2. Introduction

Adjustments, amendments and concealing of the crime evidences have resulted in an increment in the world's crime rate. One such recurring hindrance can be seen with the most vital forms of evidence; bloodstains. Bloodstains are considered to be one of the most difficult stains to remove and even after cleaning them off the clothes through the detergents available in the market, there are some colored traces left which do not disappear completely and can be detected. The red color present in the blood is due the pigment called hemoglobin which basically comprises of 'Hem which means Iron and globins which means protein.' [1] Forensic scientists use luminol to detect bloodstains which when reacted with iron in blood stains exhibit chemiluminescence. [2] This led to a thinking of till what extent were these criminals able to manage the escape by washing the stains and hinder the evidences. Thus the desire to study more about the mystery led to this investigation on bloodstains removal. After researching about the detergents from journals like '*Surface active agents: Advances in research and application-* ***Ashton*** **Acton'** and websites like '*The Science of Stain Removal, Study of Oxyper-* H_2O_2*, and Study of sodium per borate*' and also studying [3]the effect of various compounds present in them, it was found out that the active oxygen

---

[1] "Definition of Hemoglobin." *www.medicinenet.com*, May 13, 2016, https://www.medicinenet.com/script/main/art.asp?articlekey=15738. Accessed Jan 2, 2018.

[2] "What Is Chemiluminescence? | www.scienceinschool.org." *www.scienceinschool.org*, 5 undefined. 2011, http://www.scienceinschool.org/2011/issue19/chemiluminescence. Accessed Jan 2. 2018.

reagents were the sources through which bloodstain removal was possible. This helped in framing a research idea:

To what extent does the change in mass (1- 5g) of the oxidizing reagent, sodium per borate ($NaBO_3 \cdot H_2O$), sodium per carbonate ($Na_2CO_3 \cdot 1.5H_2O_2$) and hydrogen peroxide (H_2O_2)found in detergents is effective in removing bloodstains, measured by spectro-photometric analysis of change in chemiluminescence intensity of luminol due to the oxidation of ferrous (Fe^{2+}) to ferric ions (Fe^{3+}) in stock solution containing oxidizing agents and iron solution (120 µg/dL) imitating blood at room temperature (30°C)?

In this essay, the correlation between removal of the stains and the extent to which it depends on the amount of reagent was measured. The decrease in chemiluminescence signifies the extent to which the stains are removed. On this basis, the cleansing action of various oxidizing agents like hydrogen peroxide, sodium per borate and sodium per carbonate were compared. Seeing the competence of the strongest oxidizing reagent, it can be considered that this research will be useful for forensic science departments to detect the extent to which evidences were tried to be concealed as all the oxidizing agents contain H_2O_2 and even after washing the clothes, some content of H_2O_2 is left. Using peroxide test strips it can be identified whether the criminals have tried to clear the evidences or not. It will also be valuable for the detergent manufacturers because through this research, they will opt for a better oxidizing agent to yield a better cleansing action.

---

3. Background information

3.1 Oxidizing reagent

Oxidizing agents work on the phenomena of releasing oxygen molecules through the process of oxidation which helps in breaking the bonds of chemical compounds- chromophores which produce color reflecting a definite region of the visible light spectrum. [4] These oxidizing agents change the molecular shape of chromophores which leads to reflection of no color or color outside the visible spectrum. After researching from *organic chemistry portal*[5], two commonly used highly active oxidizing reagents '***Sodium per Carbonate*** ($2Na_2CO_3 \cdot 3H_2O_2$) [6] and S***odium per Borate Monohydrate*** ($NaBO_3 \cdot H_2O$) [7]' were selected as they can be easily prepared in the laboratory. ***Sodium per Carbonate*** is a colorless crystalline adduct which is denser than water. Due to its polarity, it is easily soluble in water and has the tendency to dissociate into ***Hydrogen Peroxide*** (H_2O_2) and sodium carbonate (Na_2CO_3) which together act as a complete detergent.

$$2Na_2CO_3 \cdot 3H_2O_2 \rightarrow 2Na_2CO_3 + 3H_2O + 1.5\ O_2 + Heat \uparrow$$

A long time exposure to this chemical might cause burns, skin irritation, eye irritations, blindness, throat irritation and coughing; however the amount used for households isn't harmful for humans. The long time exposure can be avoided by wearing gloves while handling the detergents.

The second reagent ***Sodium per Borate Monohydrate*** was manufactured by Deutsche Gold- und Silber-Scheideanstalt by dehydrating sodium per borate tetra-hydrate. Unlike sodium per carbonate and per phosphate, sodium per borate is not simply an adduct[8] with hydrogen peroxide. Instead, it forms a cyclic structure with a B_2O_4 core and four hydroxide groups surrounded by two boron atoms.

---

[4] "How Does Bleach Work?." *wonderopolis.org*, https://wonderopolis.org/wonder/how-does-bleach-work. Accessed Jan 3, 2018.
[5] "Oxidizing Agents." *www.organic-chemistry.org*, http://www.organic-chemistry.org/chemicals/oxidations/. Accessed Jan 3, 2018.
6"PubChem Search." pubchem.ncbi.nlm.nih.gov, https://pubchem.ncbi.nlm.nih.gov/search/. Accessed 3 Jan, 2018.
7 "Sodium Perborate - Oxygen Bleaching Chemical For Laundry Detergents." Runyoutech.com. 2017. Web. 4 Jan. 2018.
8 "Acid-Base Reaction - Dissociation Of Molecular Acids In Water | Chemistry." Encyclopedia Britannica. 2017. Web. 4 Jan 2018

$$2NaBO_2+2H_2O_2+H_2O \longrightarrow Na_2[(OH)_2B(O-O)_2B(OH)_2]\cdot H_2O$$ [9]

This makes the bond weaker, increasing the oxidizing capacity and causes the stain to be removed. A long time exposure to this may cause dermatologic manifestations[10]. This can be avoided by wearing gloves and avoiding inhalation or direct contact.

To compare the activity of the other two reagents, the third reagent chosen is ***Hydrogen Peroxide*** which is commonly used as a bleaching agent in detergents due to its excessively high oxidizing quality, easy availability and cheap rate.

$$2\,H_2O_2 \longrightarrow H_2O + O_2$$ [11]

The side effects of the reagent are that if directly used or used in excess, then it may damage the skin. To prevent this, a diluted solution of H_2O_2 should be used in the detergents.

3.2 Mechanism and Detection of Chemiluminescence

H.O. Albrecht discovered **Luminol** (technically 5-Amino-2, 3-dihydro-1, 4-phthalazinedione) [12]. When blood (containing hemoglobin) and luminol mixture (containing H_2O_2) come in contact, hemoglobin initiates an oxidation reaction between luminol and hydrogen peroxide. [13] During this reaction, luminol undergoes oxidation by gaining oxygen atoms and losing nitrogen atoms and hydrogen atoms forming an energized compound 3-aminophthalate. In it, the electrons of oxygen atoms jumps to higher orbital which when fall back into lower energy state emit photons producing blue light. This exhibits the chemiluminescence property of luminol[14].

---

9 "Preparation of Sodium Peroxoborate- Equations." *chemistry.stackexchange.com*, March 8, 2017, https://chemistry.stackexchange.com/questions/44911/preparation-of-sodium-peroxoborate-equations. Accessed Jan 6, 2018.
10 "SODIUM PERBORATE - National Library of Medicine HSDB Database." toxnet.nlm.nih.gov, May 12, 2017, https://toxnet.nlm.nih.gov/cgi-bin/sis/search/a?dbs+hsdb:@term+@DOCNO+1676. Accessed Jan 8, 2018.
11 "Hydrogen Peroxide - the Evonik History Portal - the History of Evonik Industries." history.evonik.com, http://history.evonik.com/sites/geschichte/en/inventions/hydrogen-peroxide/pages/default.aspx. Accessed Jan 8, 2018.
12 Luminol - Dictionary Definition of Luminol | Encyclopedia.com: FREE Online Dictionary." www.encyclopedia.com, Feb 25, 2018, https://www.encyclopedia.com/science/encyclopedias-almanacs-transcripts-and-maps/luminol. Accessed Jan 11, 2018.
13 Harris, Tom. "How Luminol Works." science.howstuffworks.com, June 11, 2002, https://science.howstuffworks.com/luminol2.htm. Accessed Jan 11, 2018.
14 https://eic.rsc.org/exhibition-chemistry/chemiluminescence-the-oxidation-of-luminol/2020040.article

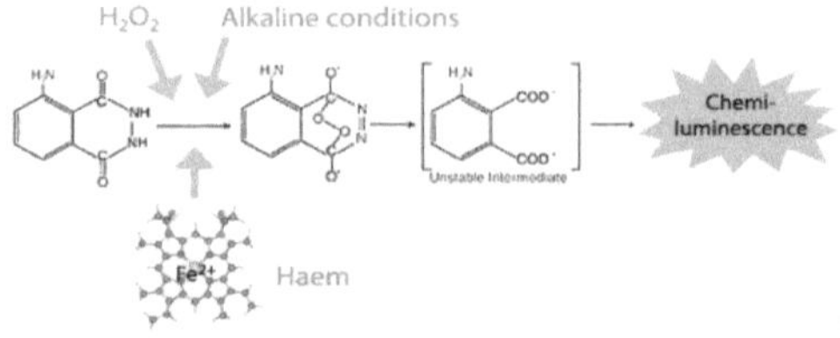

[15]

Image 1- Mechanism of Luminol Activation

The chemiluminescence of luminol can be measured through several methods like thermo-luminescence or through the light emitted by the reaction between iron and luminol in presence of H_2O_2. The method used in this experiment is **spectro-photometer** [16]which works by passing a light beam through a sample to measure the absorbance/transmittance of a sample. The wavelength emitted by luminol when in contact with iron is in the blue region i.e. 420nm- 445nm but in this experiment ***428nm of wavelength*** was used as it showed the highest transmittance during the experimentation. (Shown in the appendix- Graph 13)

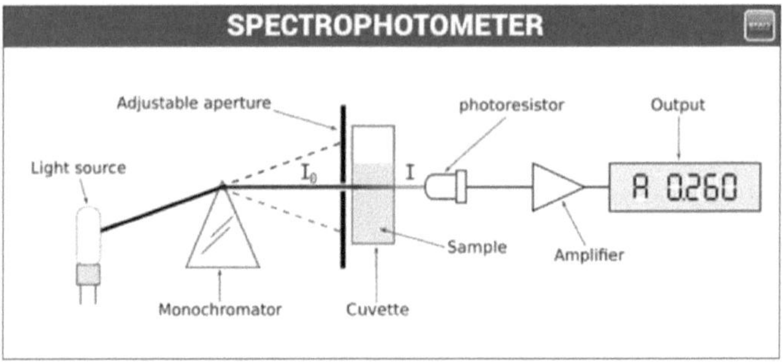

[17]

Image 2 - Working mechanism of spectrometer

---

[15] "BLUESTAR Forensic - Chemistry and History." *www.bluestar-forensic.com*, http://www.bluestar-forensic.com/gb/bluestar-chemistry.php. Accessed 12 Jan. 2018.

[16],[16] "Principle of Spectrophotometer and Its Applications| Chemistry| Byjus." *byjus.com*, Jan 20, 2017, https://byjus.com/chemistry/spectrophotometer-principle/. Accessed Jan 11, 2018.

4. Research Variables

The amount of sodium per borate, hydrogen peroxide and sodium per carbonate was chosen as the ***Independent Variable*** as varying the amount of reagent used would show different extents of the activity of oxidizing agent on bloodstain stain removal. The ***absorbance pattern emitted by the usage of oxidizing agent*** was chosen as the ***Dependent Variable*** as the absorbance pattern shown at the 428nm of wavelength by the all three reagents depends on the mass of reagent used. This was measured by the spectro-photometric analysis of luminol. The ***amount of sodium silicate and sodium carbonate*** used was kept ***Controlled*** as Sodium silicate moves in prevention of deposition of mineral at the surface resulting in hardness and sodium carbonate was used to avoid bonding of hard water with the detergent. The two chemicals were used as a complimentary detergent ingredient to obtain better results and hence had to be constant for all the trials. ***Amount of luminol stock solution and iron stock solution*** used were chosen as the ***Controlled Variable*** as luminol helps in detection of blood stains and it was taken in proportionate amount to iron stock solution. More the iron solution, more detergent will be required to remove the stain which will result in hindrance of the results which is why it was necessary to control both the quantities. The third ***controlled variable*** was ***temperature*** which was kept at 30°C because changing the temperature of solutions will change the cleansing action is detergents

5. Expectation

When luminol reacts with bloodstains (specifically Fe^{2+} (ferrous ions in blood)), it emits a blue light indicating the presence of blood. Blood has red color due to hemoglobin and the oxidizing agents present in detergents easily break the bonds present between the chromophores. Chromophores have an ability to possess color. Due to breaking of the bonds, they will emit light which does not have a wavelength in the visible light spectrum. So using the above stated phenomena, the experiment was conducted to demonstrate the effect of oxidizing reagents on removal of the red pigment from the bloodstains. It is assumed that when iron (Fe^{2+}) present in hemoglobin reacts with the oxidizing reagent, they get oxidized into ferric ions (Fe^{3+}). Ferrous ions have the ability to bind with luminol producing chemiluminescence which helps in detecting bloodstains whereas the oxidized form ferric ions can no more bind with luminol. It implies that the oxidizing strength of the reagent is directly proportional to the ability of the reagent to work on stain removal. This change in oxidizing strength can be measured by the absorbance spectrum obtained by photo-spectroscopy of a particular wavelength (428nm) in visible region.

6. Chemicals required

Table 1 Chemicals Required[18]

Serial Number	Chemical name	Company's name	Amount of chemical used
1	Mohr's Salt ($(NH_4)_2Fe(SO_4)_2.6H_2O)$)	Rankem	0.84g
2	Luminol ($C_8H_7N_3O_2$)	Ozone International	0.005g
3	Anhydrous sodium carbonate (Na_2CO_3)	Rankem	0.2g
4	Sodium Bicarbonate ($NaHCO_3$)	Rankem	1.2g
5	Distilled Water	-	3litre
6	Hydrogen Peroxide (H_2O_2)	Merck	40 cm^3
7	Borax Powder ($Na_2[B_4O_5(OH)_4].8H_2O$)	Rankem	24g
8	Sodium per carbonate ($Na_2CO_3.1.5H_2O_2$)	Sigma-Aldrich	6g
9	Sulfuric acid (H_2SO_4)	Merck	10 cm^3
10	Ethyl Ether ($(C_2H_5)_2O$)	Rankem	25 cm^3
11	Ethyl Alcohol (C_2H_6O)	Changshu Yanguan	25 cm^3
12	Hydrated ammonium carbonate ($CH_{10}N_2O_4$)	Merck	0.0025g
13	Sodium Silicate (Na_2SiO_3)	ASES chemical works	20g
14	Sodium Hydroxide (NaOH)	Rankem	5g

[18] 19 Author's own work

7. Apparatus required

Table 2 Apparatus Required[19]

Serial number	Apparatus used		Quantity
1	Beaker 100cm^3		10
2	Test tube		2
3	Conical flask 250cm^3		2
4	Dropper		5
5	Tumbler		1
6	Water bath		1
7	Tong		3
8	Filter paper		5
9	Blotting paper		1
10	Stirrer		2
11	Weighing balance		1
12	Watch glass		5
13	Spatula		5
14	Pipette 1 cm^3		1
15	Measuring cylinder	1) 50 cm^3	2
		2) 100 cm3	1

8. Cautionary Principles

Before beginning with the experimentation, safety glasses and lab coat were worn to protect the eyes and body. All the measurements were read perpendicularly in order to avoid parallax error and the weighing balance was pre calibrated to avoid zero error. For making monohydrate from tetra hydrate, blotting paper was used to remove excess water content from the crystals before putting it for the water bath. It was made sure that the chemicals and oxidizing reagents were left for 5 minutes in the cuvette after vigorous shaking in order to avoid sedimentation of the chemicals. To store the chemicals in the glass tubing, lubricants were used to fit the corks as pressuring the stoppers can lead to breakage. [20] To avoid the cap of the cuvette from popping due to the pressure created by the oxidizing agents, it was let loose after shaking so that the gas escapes. Spills of these reagents were be contained and isolated from waterways and sewer drains. Any remaining residues were diluted with large amounts of water.

---

[20] "." *www.solvay.us*, https://www.solvay.us/en/binaries/PSS-Sodium-Percarbonate-164361.pdf. Accessed Nov 16, 2017.

9. Methodology of Experimentation:

9.1 Preparations of the chemicals

Synthesizing of luminol stock solution:

As per the study conducted by ***Sigma Aldrich***, luminol has a water soluble property and requires a catalyst (Ferrous ions) for its activation. [21] The research paper of ***Carolina biological study*** stated that to prepare a luminol stock solution, 0.005g of luminol, 0.2g anhydrous sodium carbonate (Na_2CO_3), 1.2g of sodium bicarbonate ($NaHCO_3$), and 0.0025g of hydrated ammonium carbonate [$(NH_4)_2CO_3H_2O$] were mixed together[22]. To this mixture, 50cm^3 of distilled water was added and stirred for over 2 minutes. A few drops of H_2O_2 were mixed to activate the luminol solution.

Picture 1.1 (preparation of luminol stock solution)[23]

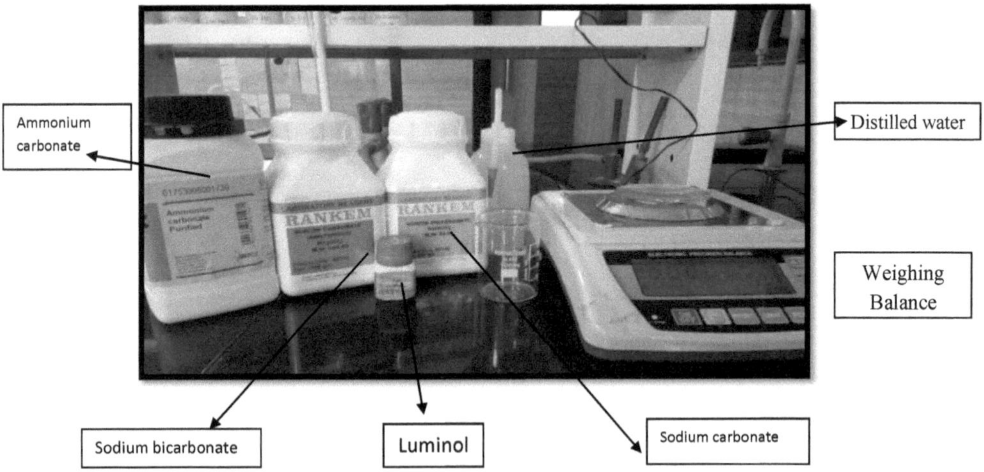

[21] "." *www.sigmaaldrich.com*, https://www.sigmaaldrich.com/content/dam/sigma-aldrich/docs/Sigma/Product_Information_Sheet/a8511pis.pdf. Accessed Nov 17, 2017.

[22] "How to Make Luminol Glow: The Glowing Reaction." *www.carolina.com*, https://www.carolina.com/teacher-resources/Interactive/luminol-glowing-reaction/tr10786.tr. Accessed Nov 17, 2017.

[23] Author's own work

Preparation of Iron stock solution imitating blood:

The usual range of iron in blood is [24]60-170 μg/**dL**. As hemoglobin value differs in different people according to their age, sex and nutrition, an averaged concentration of 120 μg/**dL** of iron was chosen. Iron solution was obtained through Mohr's salt $(NH_4)_2Fe(SO_4)_2 \cdot 6H_2O$. Molar mass of Mohr's salt = 392 g/mol. So,

$$
\begin{aligned}
&1mol \text{ mohr's salt} = 1 \text{ mol of } Fe^{2+} \\
&392\,g/mol = 56g\ of\ iron \\
&X \text{ g/mol} = 0.12g\ of\ iron \\
&X = 0.84g\ of\ Mohr's\ salt
\end{aligned}
$$

8.4mg of Mohr's salt was added to a little amount of H_2SO_4 (4-5 drops) in order to prevent hydrolysis of Mohr's salt. This solution was further diluted with distilled water to form 1 liter of iron stock solution.

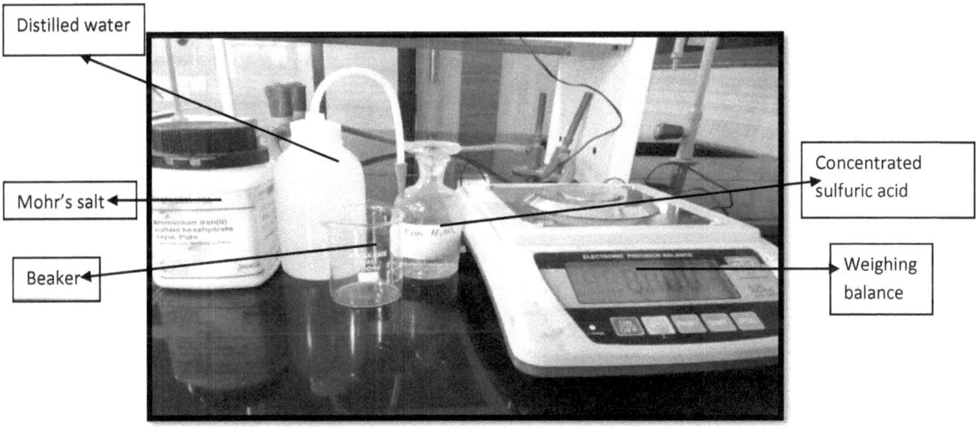

Picture 1.2 (preparation of iron stock solution)[25]

---

[24] "Serum Iron Test: Purpose, Procedure, and Results." *www.healthline.com*, June 1, 2012, https://www.healthline.com/health/serum-iron#uses. Accessed Feb 26, 2018.

[25] Author's own work

Synthesis of sodium per borate monohydrate:

Sodium per borate monohydrate is a strong oxidizing reagent ($4Na_2BO_3+5H_2O$). 24g of borax powder and 5g of sodium hydroxide was mixed with 150cm^3 of distilled water heated up to 45°C.[26] The solution was left for 10 minutes to reach the room temperature. 28.3cm^3 of 30% of H_2O_2 solution was diluted in 254.7cm^3 of distilled water. After a vigorous stirring, H_2O_2 was added to the beaker containing the solution. The beaker became hot due to the exothermic reaction. To lower the temperature, the beaker was placed in the middle of a large container with ice for constant cooling. After a few minutes, few crystals of sodium per-borate started separating. The solution was stirred for next twenty minutes continuously to make fine crystals which were separated out. The crystals were washed with 2 successive portions of 25cm^3 of diethyl ether and 25cm^3 of ethyl alcohol to remove other organic impurities and were stored in a stopper bottle. Through this method tetra hydrate was made which when heated at 70°C on digital water bath formed sodium per borate mono-hydrate.

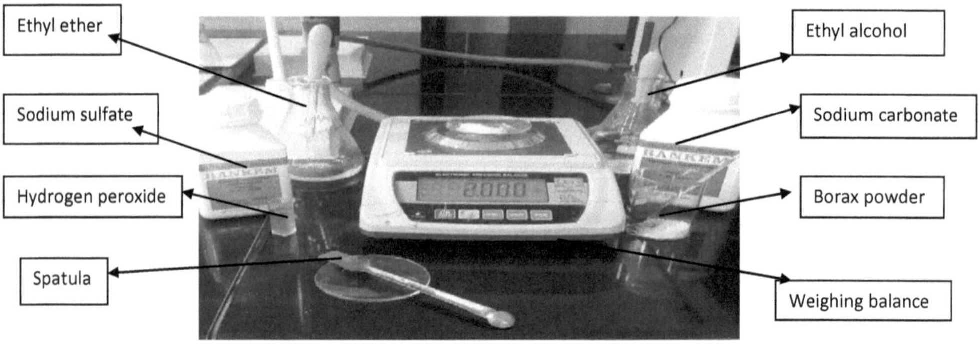

Picture 1.3 (preparation of sodium per borate)[27]

[26] Blanchard, A. A. "Synthesis of SODIUM PERBORATE - PrepChem.com." www.prepchem.com, Sept 19, 2015, http://www.prepchem.com/synthesis-of-sodium-perborate/. Accessed 20 Nov, 2017.

[27] Author's own work

Preparation of 3% H_2O_2:

As standardized concentration of H_2O_2 content in any household compound is 3%, a diluted solution of H_2O_2 was to be formed due to the availability of 30% concentrated solution in the laboratory[28]. It is highly reactive and can harm the skin. To prevent this, gloves and goggles were worn. Now for the conversion:

$$
\begin{aligned}
&3\% = 10ml \\
&30\%\, is\, 10 \text{ times of } 3\% \\
&\text{So, } \frac{10}{10} = 1ml\, of\, H_2O_2 \\
&Giving, 1ml\, of\, H_2O_2\, added\, to\, 9ml\, of\, distilled\, water
\end{aligned}
$$

The solution was stirred for 2 minutes and was stored in a stopper bottle to avoid the escape of activated oxygen gas.

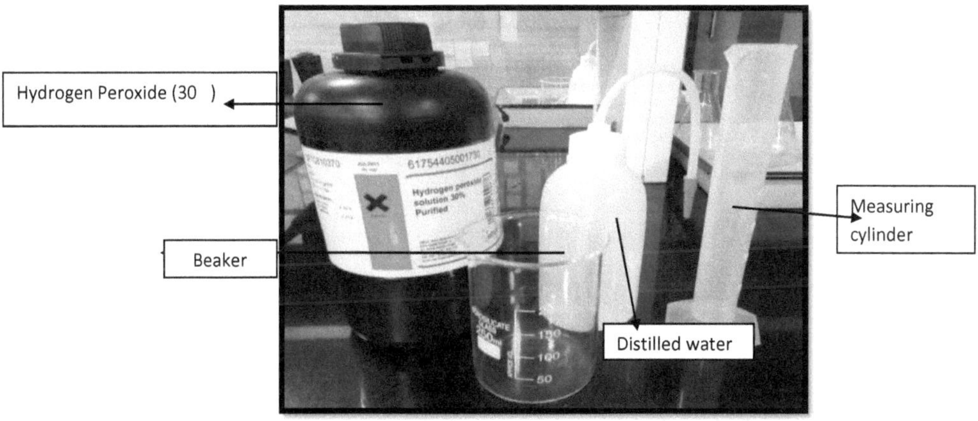

Picture 1.4 (Preparation of hydrogen peroxide)[29]

[28] "Hydrogen Peroxide Advanced Oxidation Processes." *Spartanwatertreatment.com*. 2017. Web. 21 Nov. 2017.

[29] Author's own work

9.2 Using the spectrophotometer

Set up of the experiment

Once the chemicals were made (Iron solution, luminol solution and the three reagents: H_2O_2, Sodium per borate and sodium per carbonate), they were to be put into the spectrophotometer to check their action. 0.5g of sodium carbonate (a common detergent found in the washing powder) was taken in beaker and mixed with 0.2g of sodium silicate to form a base of the detergent. [30] Now for the first trial, 0.1g of sodium per-borate monohydrate was added and mixed vigorously with the previously formed mixture (Sodium carbonate + sodium silicate). A blank cuvette with 4cm^3 of distilled water was added to the spectrometer to calibrate it and then using a 1cm^3 pipette, 0.5cm^3 of luminol stock solution and 1cm^3 of iron stock solution was put into the cuvette of the spectrophotometer. A sudden blue glow was observed in the solution which indicated that luminol was activated and was indicating the presence of iron.

Collection of the data

After waiting for 5 minutes for the chemicals to mix well with each-other, all three chemicals (sodium carbonate, sodium silicate and sodium per borate) were added to the cuvette. In Lab Quest 2, 'Intensity' option was chosen and the absorption wavelength was set to 428nm from the entire wavelength ranging from 300nm-700nm. The spectrometer was run for 60 seconds and then the data was saved. 5 trials were collected for each mass (0.1-0.5g) of all three oxidizing agents (sodium per borate, sodium per carbonate and hydrogen peroxide.) Spectrophotometer was calibrated by using cuvette with distilled water between every 2 trials to reset the device and avoid uncertainty. After every experiment, the droppers, cuvette and measuring cylinders were washed to guarantee that the results weren't hindered by any of the chemical reaction. After the experimental data was collected in Lab Quest 2, it was transferred to the Logger Pro application. Using the data collected and observing the graphs, processed data tables were formed.

10. Qualitative analysis

Luminol is a fine white powder which when mixed with ferrous iron stock solution imitating blood yielded an instant blue spark indicating the presence of ferrous ions. When sodium silicate and sodium carbonate were added to this solution, the solution turned milky. Later when oxidizing agents are added to this mixture; a fizzy sound was produced with effervescence of oxygen exerting pressure on the cap.

11. Data collection and Graphical analysis

- Sample calculations:

- **Average calculations**

$$\frac{Absorbance_1 + Absorbance_2 + Absorbance_3 + Absorbance_4 + Absorbance_5}{5} = \text{Average Absorbance}$$

$$\frac{2.023+2.294+2.115+2.029+2.002}{5} = 2.093 \text{ arb. units}$$

- **Standard deviation**

$$\sqrt{\frac{\sum (x-\bar{x})^2}{n}} = \sigma$$

$$\sqrt{\frac{\sum (2.023-\mathbf{2.093})^2}{5}} = \mathbf{0.120}$$

- **Uncertainty calculation:**

$$\frac{Maximum\ value\ of\ the\ absorbance - Minimum\ value\ of\ absorbance}{2}$$

$$= maximum\ uncertainty$$

$$\frac{2.294 - 2.002}{2} = \pm 0.150$$

- The same was done for the data and data tables to calculate the average, standard deviation and the uncertainty

'Compound' refers to sodium per borate, sodium per carbonate and hydrogen peroxide

11.1 Absorbance pattern shown by 0.1g of the reagents in visible light region

- Graph-1.1- Absorbance pattern shown by 0.1g of sodium per borate in visible light region[31]

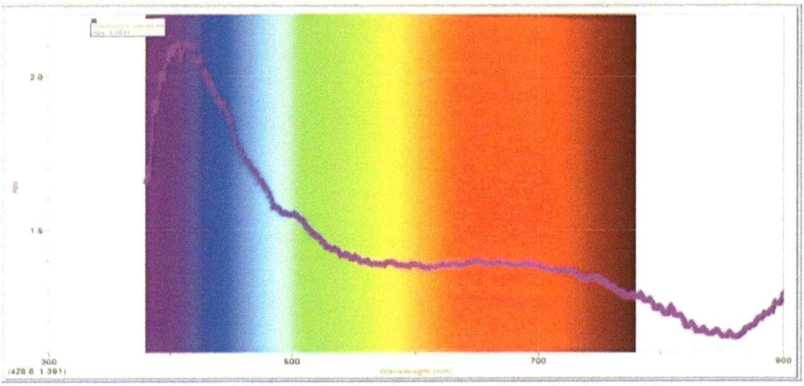

[31] Author's own work

- Graph-1.2 Absorbance pattern shown by 0.1g of hydrogen peroxide in visible light region[32]

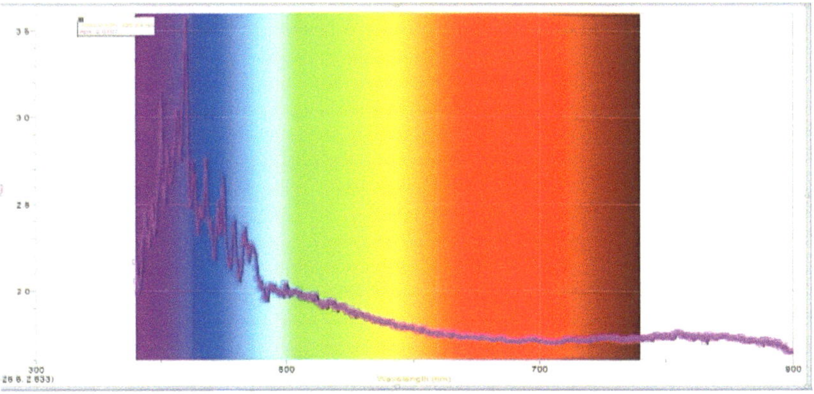

- Graph-1.3- Absorbance pattern shown by 0.1g of sodium per carbonate in visible light region[33]

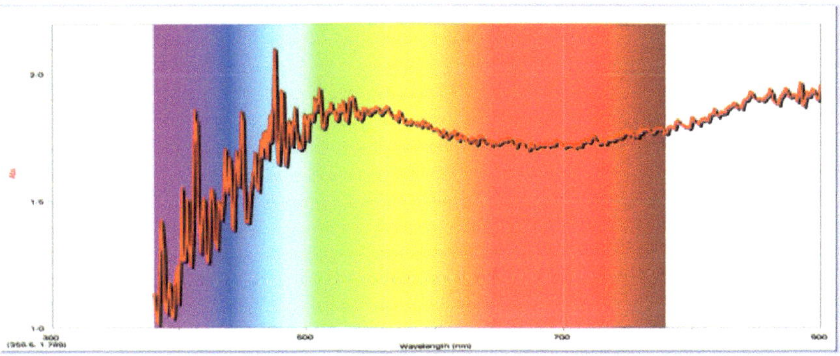

[32] Author's own work

[33] Author's own work

- Table 3 - 0.1g of the compound used and the absorbance shown in arbitrary units[34]

Number of trials	***Mass of the compound (g ± 0.001g)***	***Sodium per borate (arbitrary unit)***	***Sodium per carbonate (arbitrary unit)***	***Hydrogen peroxide (arbitrary unit)***
1	0.1	2.023	1.454	2.001
2	0.1	2.294	1.227	2.045
3	0.1	2.115	1.447	1.995
4	0.1	2.029	1.300	1.822
5	0.1	2.002	1.325	2.017
Average absorbance		**2.093**	**1.351**	**1.933**
Standard deviation		**0.120**	**0.100**	**0.090**
Uncertainty calculation (±)		**0.150**	**0.110**	**0.120**

11.2 Absorbance pattern shown by 0.2g of the reagents in visible light region

- Graph-2.1- Absorbance pattern shown by 0.2g of sodium per borate in visible light[35]

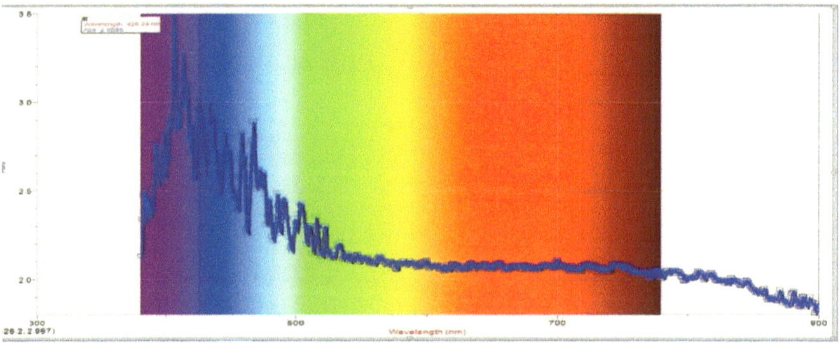

[34] Author's own work
[35] Author's own work

- Graph-2.2- Absorbance pattern shown by 0.2g of hydrogen peroxide in visible light region[36]

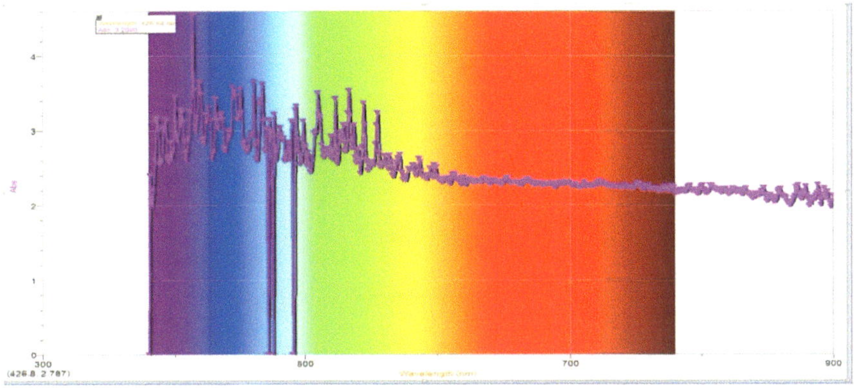

- Graph-2.3 - Absorbance pattern shown by 0.2g of sodium per carbonate in visible light region[37]

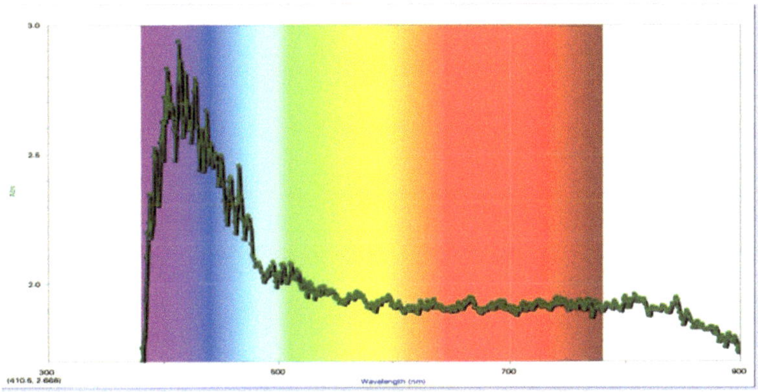

36 Author's own work
37 Author's own work

- Table 4 - 0.2g of the compound used and the absorbance shown in arbitrary units[38]

Number of trials	*Mass of the compound (g ± 0.001g)*	***Sodium per borate (arbitrary unit)***	***Sodium per carbonate (arbitrary unit)***	***Hydrogen peroxide (arbitrary unit)***
1	0.2	2.666	2.610	2.631
2	0.2	2.577	2.569	2.570
3	0.2	2.642	2.522	2.532
4	0.2	2.633	2.469	2.623
5	0.2	2.684	2.516	2.656
Average absorbance		**2.598**	**2.537**	**2.573**
Standard deviation		**0.040**	**0.050**	**0.050**
Uncertainty calculation (±)		**0.050**	**0.070**	**0.060**

[38] Author's own work

11.3 Absorbance pattern shown by 0.3g of the reagents in visible light region

- **Graph-3.1 Absorbance pattern shown by 0.3g of sodium per borate in visible light region[39]**

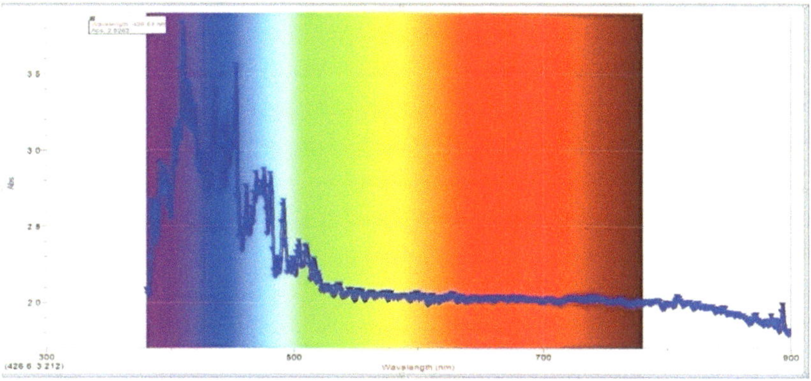

- **Graph-3.2 Absorbance pattern shown by 0.3g of hydrogen peroxide in visible light region[40]**

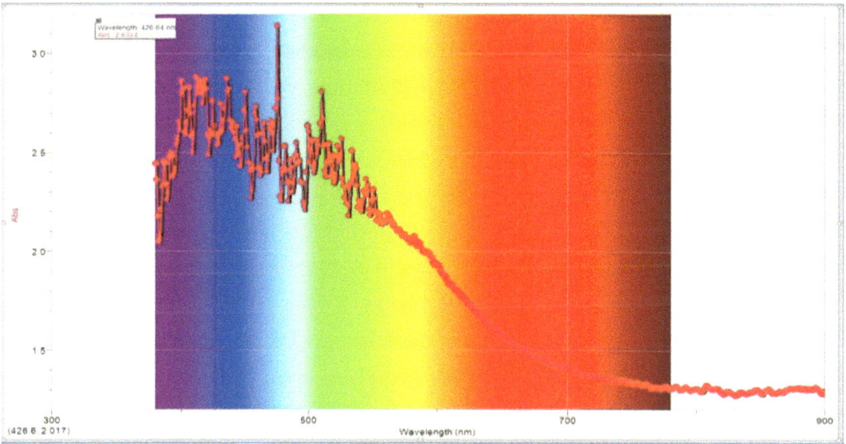

[39] Author's own work
[40] Author's own work

- Graph 3.3 Absorbance pattern shown by 0.3g of sodium per carbonate in visible light region[41]

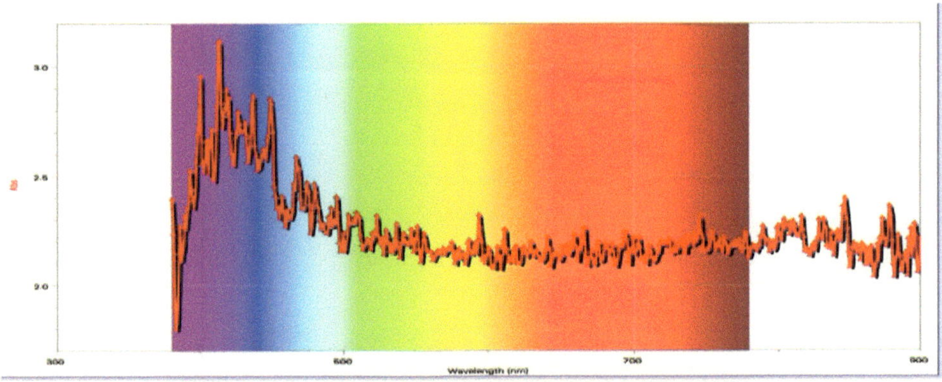

- Table 5- 0.3g of the compound used and the absorbance shown in arbitrary units[42]

Number of trials	*Mass of the compound (g ± 0.001g)*	*sodium per borate (arbitrary unit)*	*Sodium per carbonate (arbitrary unit)*	*Hydrogen peroxide (arbitrary unit)*
1	0.3	2.983	2.699	2.819
2	0.3	2.764	2.686	2.725
3	0.3	2.742	2.672	2.732
4	0.3	2.833	2.583	2.623
5	0.3	2.784	2.661	2.756
Average absorbance		**2.821**	**2.660**	**2.671**
Standard deviation		**0.100**	**0.050**	**0.070**
Uncertainty calculation (±)		**0.120**	**0.060**	**0.100**

[41] Author's own work
[42] Author's own work

11.4 Absorbance pattern shown by 0.4g of the reagents in visible light region

- **Graph-4.1 Absorbance pattern shown by 0.4g of sodium per borate in visible light region[43]**

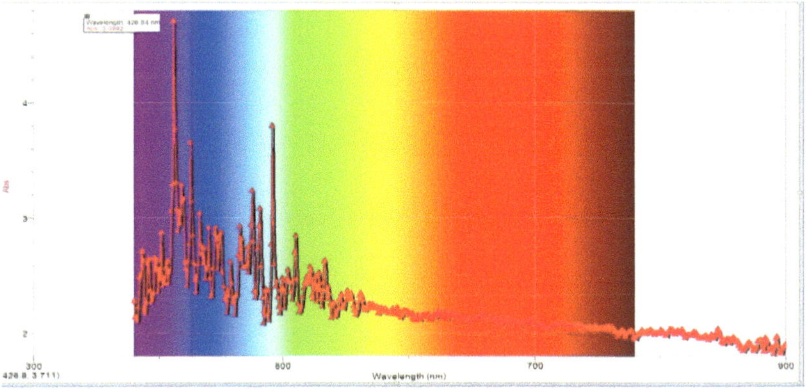

- **Graph-4.2 Absorbance pattern shown by 0.4g of hydrogen peroxide in visible light region[44]**

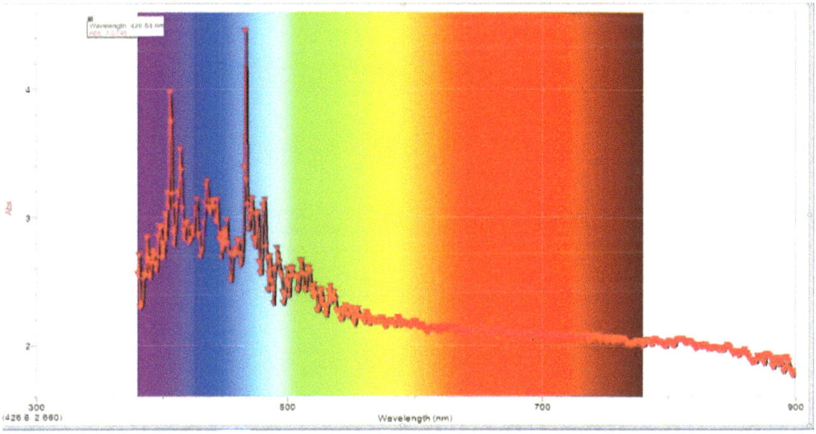

[43] Author's own work
[44] Author's own work

- **Graph 4.3 Absorbance pattern shown by 0.4g of sodium per carbonate in visible light region**[45]

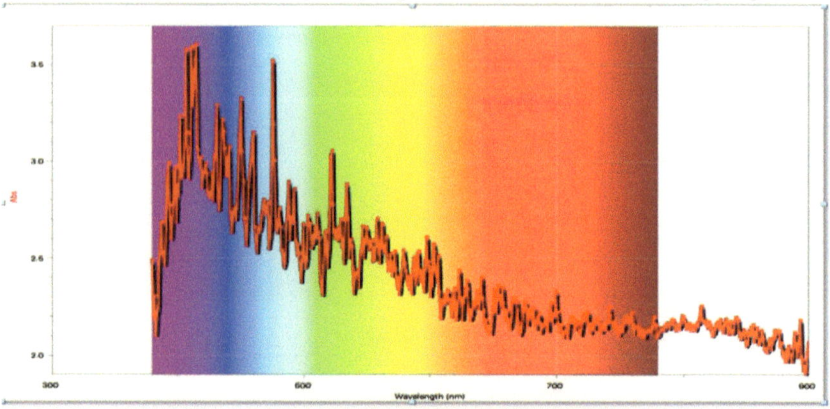

- **Table 6- 0.4g of the compound used and the absorbance shown in arbitrary units**[46]

Number of trials	*Mass of the compound (g ± 0.001g)*	***Sodium per borate (arbitrary unit)***	***Sodium per carbonate (arbitrary unit)***	***Hydrogen peroxide (arbitrary unit)***
1	0.4	3.011	2.799	2.857
2	0.4	2.998	2.781	2.725
3	0.4	3.105	2.725	2.832
4	0.4	2.889	2.751	2.823
5	0.4	2.984	2.549	2.756
Average absorbance		**2.997**	**2.721**	**2.798**
Standard deviation		**0.080**	**0.100**	**0.060**
Uncertainty calculation (±)		**0.110**	**0.130**	**0.070**

[45] Author's own work
[46] Author's own work

11.5 Absorbance pattern shown by 0.5g of the reagents in visible light region

- **Graph-5.1 Absorbance pattern shown by 0.5g of sodium per borate in visible light region**[47]

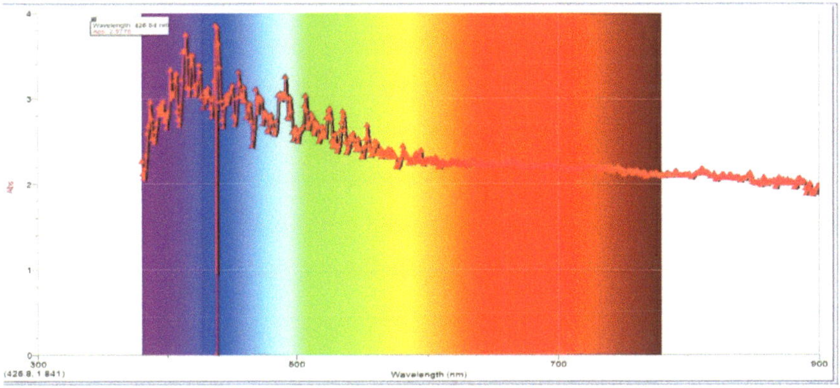

- **Graph-5.2 Absorbance pattern shown by 0.5g of hydrogen peroxide in visible light region**[48]

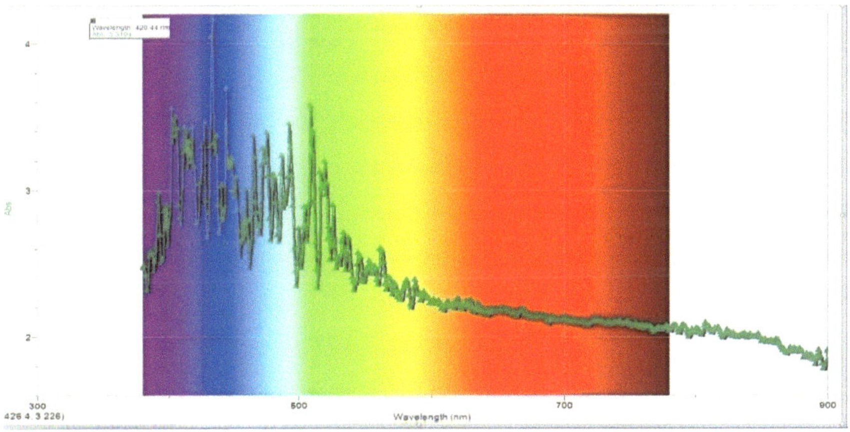

[47] Author's own work
[48] Author's own work

- **Graph 5.3 Absorbance pattern shown by 0.5g of sodium per carbonate in visible light region[49]**

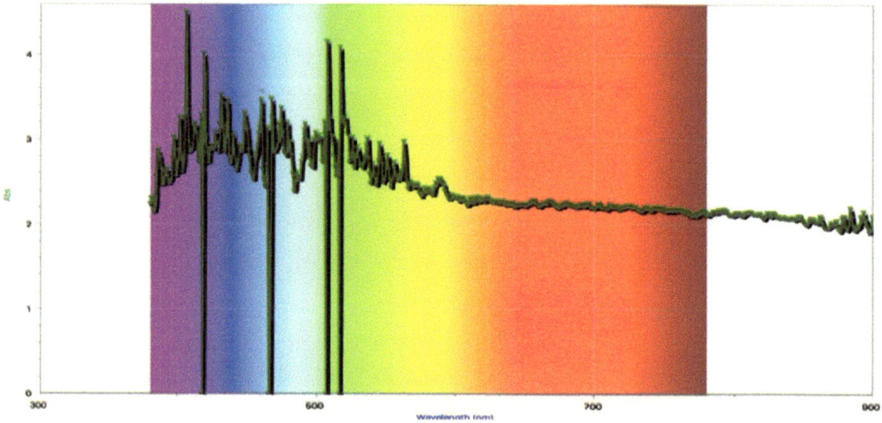

- **Table 7 - 0.5g of the compound used and the absorbance shown in arbitrary units[50]**

Number of trials	*Mass of the compound* (g ± 0.001g)	*Sodium per borate* (arbitrary unit)	*Sodium per carbonate* (arbitrary unit)	*Hydrogen peroxide* (arbitrary unit)
1	0.5	3.463	2.844	2.930
2	0.5	3.367	2.862	2.825
3	0.5	3.454	2.811	2.932
4	0.5	3.290	2.865	2.923
5	0.5	3.222	2.859	2.856
Average absorbance		**3.359**	**2.848**	**2.893**
Standard Deviation		**0.100**	**0.020**	**0.050**
Uncertainty calculation (±)		**0.120**	**0.030**	**0.050**

[49] Author's own work
[50] Author's own work

12. Explanation of results and discussion

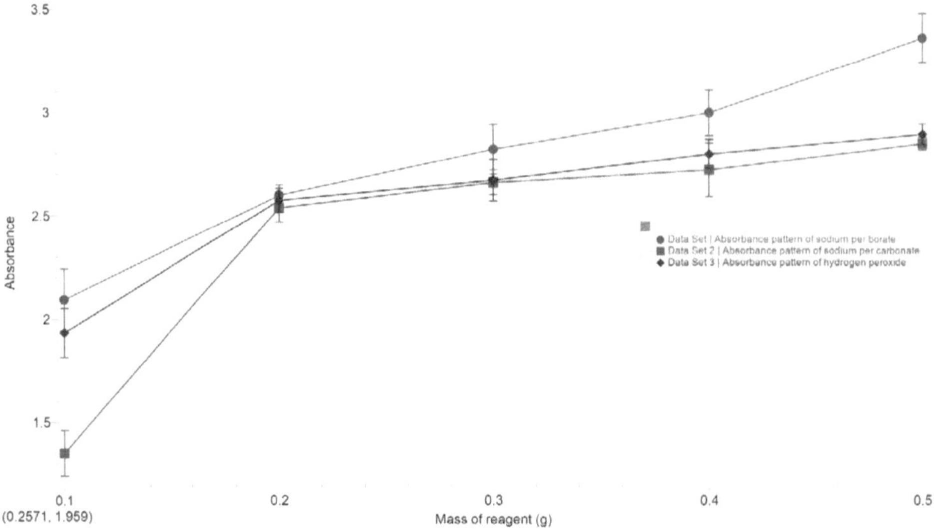

Graph 6 (Comparison between mass of reagent used and the absorbance pattern shown by all three reagents)[51]

The results obtained experimentally through graph 13 in the appendix, show that when luminol and iron solution were analyzed in the spectrometer, the highest wavelength transmitted was ranging from 426-430nm however, as the oxidizing reagent was added to the solution, exact opposite happened i.e. the highest absorbance was shown by the wavelength between 426-430nm which lies in the blue region of the spectrum. This is due to the chemiluminescence property of luminol produced when it reacts with ferrous ions and oxidizes it into ferric ions. This increases the transmittance of blue light corresponding to the wavelength of 428nm which was used as a standard throughout the experiment. Furthermore, as the ferrous ions get oxidized to ferric ions due to oxidizing

[51] Author's own work

eagents found in detergents, they cannot react with luminol anymore reducing its chemiluminescence and hence the transmittance decreases and absorbance at 428nm wavelength of light increases.

Table 8 (Comparative analysis of mass of the reagent used and the absorbance shown by the reagents)[52]

Mass of the reagent (g)	Absorbance shown by sodium per borate (arb. Unit)	Absorbance shown by sodium per carbonate (arb. Unit)	Absorbance shown by H_2O_2 (arb. Unit)
0.1	2.093	1.351	1.933
0.2	2.598	2.537	2.573
0.3	2.821	2.66	2.671
0.4	2.997	2.721	2.798
0.5	3.359	2.848	2.893

While analyzing graph 6 and table 8, it can be clearly seen that there is a positive, linear trend. For sodium per carbonate and hydrogen per oxide the increment in the absorbance is higher from 0.1g to 0.3g whereas as the mass further increases, the difference in the absorbance isn't as significant as that of sodium per borate as sodium per carbonate dissociates into sodium carbonate and H_2O_2 which acts as a complete cleanser in its own. However, for sodium per borate the increment in absorbance becomes significant as the mass further increases from 0.3 to 0.5g as the amount of H_2O_2 used to make sodium per borate is lesser than that used to make sodium per carbonate. So as the mass increases more H_2O_2 is present in sodium per borate and hence the cleansing action increases. When oxidizing reagents of 0.1g mass are used, ***Sodium per Borate*** exhibits an absorbance of 2.093 arb. unit, ***Sodium per Carbonate*** of 1.351 arb. unit and ***Hydrogen Peroxide*** of 1.933 arb. unit. As the mass of the reagent is increased to 0.2g, the absorbance shown by ***Sodium per Borate*** increases to 2.598 arb. unit which is increased by 0.505 arbitrary units, for ***Sodium per Carbonate*** it increases to 2.53 arb. unit which is increased by 1.186 arb.

[52] Author's own work

units and for ***Hydrogen Peroxide*** it increases to 2.753 arb. unit which is increased by 1.28 arb. units. ***This states that more the amount of oxidizing agent is used, more Fe^{2+} ions present in the iron stock solution imitating blood are being oxidized into Fe^{3+} ions which reduce the chemiluminescence of luminol increasing the absorbance of the blue light.***

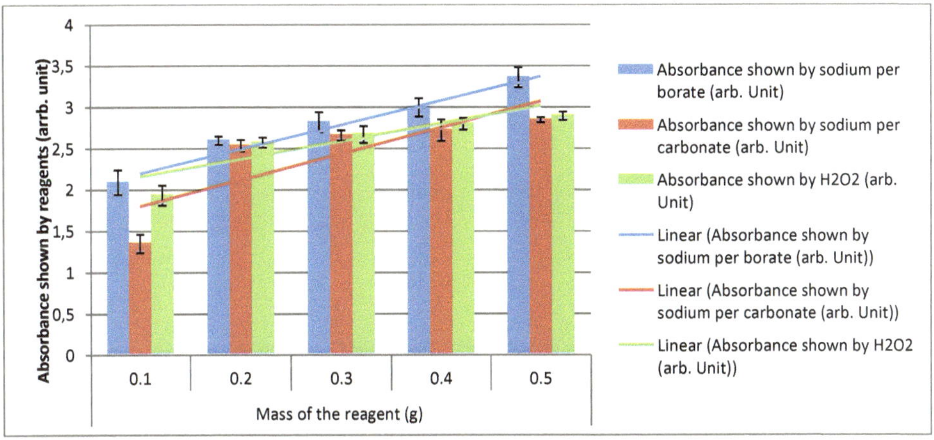

Graph 7- Comparative analysis of mass of the reagent used and the absorbance shown by the reagents[53]

[53] Author's own work

Furthermore, Pearson correlation coefficient test was carried out between the increment in the mass of reagent used and the absorbance shown by the reagent. The increment in the absorbance according to the mass is shown through the correlation as follows:

Table 9 (Increment in the absorbance shown by each reagent and the correlation between the mass and absorbance)[54]

Mass of reagent used	Increment in absorbance of sodium per borate	Increment in absorbance of sodium per carbonate	Increment in absorbance of hydrogen per oxide
0.1	0.000	0.000	0.000
0.2	0.505	1.186	0.64
0.3	0.223	0.123	0.098
0.4	0.176	0.061	0.127
0.5	0.362	0.127	0.095
Correlation	0.983	0.824	0.897

It shows that for all the experiments done for increment in mass, a positive correlation coefficient is obtained. For ***Sodium per Borate*** the correlation is the highest which is 0.983arb. unit showing that the increment in the results is highest as it has 16% active oxygen and it easily decomposes into water and oxygen[55]. The second strongest is ***Hydrogen Peroxide*** with a correlation of 0.897arb. unit, showing that it has better activity than ***Sodium per Carbonate*** which has least correlation of 0.824 arb. unit. This is so because Hydrogen Peroxide has less stability due to its O-O bonds hence it undergoes chemical reaction to form O_2 which makes it

[54] Author's own work

[55] Pesman E., Kalyoncu E. E., Kirci H.; "Sodium Perborate Usage instead of Hydrogen Peroxide for the Reinforcement of Oxygen Delignification." FIBRES & TEXTILES in Eastern Europe (2010), Vol. 18, No. 6 (83). Web. Nov 21,2017

thermodynamically stable. [56] Sodium per carbonate has least oxidizing nature due to its ability to react with reducing agents to generate unwanted heat and other gaseous products. [57]

13. Evaluation

The experiment carried out was planned to find the oxidizing strength of the chemicals that are used to remove the bloodstains at crime sites. It was first conducted with only oxidizing agents who yielded extremely high values of absorption signaling high extent of oxidation which were unclear to make any conclusion. Having high values of oxidation can be harmful for clothes as they can oxidize the fabric bonds and cause bleaching/ discoloration of clothes. Later, after referring to a study by ***Accord on 'laundry detergent ingredients'***, it was found out there are compounds used in detergents which help in better cleansing action. Those two complimentary compounds were sodium silicate and sodium carbonate. Sodium silicate helps in prevention of sedimentation of the minerals and sodium carbonate helps in avoiding bonds between hard water and the stains which when compared to the literature results, yielded the most credible results. This is why a mixture of oxidizing agents, sodium silicate and sodium carbonate was used for this research. Through the spectroscopy, which is a best method of finding the absorption pattern of a component, the peak wavelength reflected by luminol was identified. More negative the value of absorption, more the action of Luminol with iron solution. The peak value turned out to be ranging from 426-430nm as luminol emits light in blue region. It was observed that the absorption value of the spectrometer started turning positive as the mass of the oxidizing agents was further increased. ***The end results were that Sodium per Borate had the best oxidizing strength then was of Hydrogen Peroxide and then the least was of Sodium per Carbonate***. It can also be concluded that sodium per borate has the best activity with its complimentary chemicals-sodium silicate and sodium carbonate which makes

[56] "meritnation.com." *www.meritnation.com*, https://www.meritnation.com/ask-answer/question/why-h2o2-is-better-oxidizing-agent-than-water-is-there-anyth/hydrogen/6828279. Accessed Nov 21, 2017.

[57] "Sodium Percarbonate." *www.organic-chemistry.org*, http://www.organic-chemistry.org/chemicals/oxidations/sodiumpercarbonate.shtm. Accessed Nov 22, 2017.

it most suitable to be used with other components of detergents. These results were compared with the study conducted by ***Emrah Pesman, Evren E. Kalyoncu, and Hüseyin Kirci on 'Sodium Per borate Usage instead of Hydrogen Peroxide for the Reinforcement of Oxygen Delignification.'***

14. Conclusion

As per the analysis, the best oxidizing agent used in the method is ***Sodium per Borate*** which oxidized ferrous ions (Fe^{2+}) into ferric ions (fe^{3+}) fastest, increasing the absorption of the light in 428nm wavelength. As suggested in the table 8, page 27, it was noticed that more the amount of oxidizing agent found in detergents, more was the average absorption by each reagents. This showcases that increasing the mass, increases the effectiveness of bloodstains removal and decreases the chemiluminescence exhibited by luminol. This confirms the model expected in the hypothesis and the research question stating that **increasing mass of oxidizing reagents from 0.1-0.5g is effective in removing bloodstains shown by decrease in chemiluminescence property of luminol measured by spectrophotometer**. This happens because of the properties of the oxidizing agents which break the chemical bonds present in the chromophores by changing the molecular shape of the chromophores by releasing oxygen and converting them into compound which cannot possess color. The removal of this color hence proves that the oxidizing agent is suitable for the removal of bloodstains. As the mass of reagents are gradually increasing, the absorbance pattern shown by them is also increasing. ***The oxidizing property of Sodium per Borate is the highest and then is of Hydrogen Peroxide and the least is of Sodium per Carbonate.***

15. Limitations and improvements

1) Sodium per borate monohydrate was made in the laboratory from tetra hydrate, so the content of water left after boiling on the water bath couldn't be ensured. However, it can be improved by extending the time taken to boil the solution by 10-20 minutes. This should be done by controlling the temperature so that the crystals do not get dehydrated or decomposed.
2) Wavelength at which Luminol's intensity was measured was averaged to 428nm as the range was from 426-230nm, so the wavelength at which best results were given was averaged in order to maintain a consistency in the results. It can be improved by taking results for a range of different wavelengths and then taking a mean of all the absorption patterns, but this requires a large time duration and calculation,
3) The iron solution imitating blood only contained iron as the key component, as blood is a mixture of many components so the exact reaction with other components of blood couldn't be checked. It can be improved by carrying out the experiment with true blood at a more professional level with safety precautions.

16. Future scope and Extension

The experiment will be a great help for the forensic department as all the oxidizing agents contain H_2O_2 and even after washing the clothes some content of H_2O_2 is left so with the help off peroxide test strips which can be used to investigate whether the criminals have tried to clear the evidences or not. Investigators may use other chemicals to detect bloodstains like fluorescein instead of luminol which also show chemiluminescence but with different mechanism. Other oxidizing agents like sodium hypochlorite, potassium dichromate, etc can be used. Other than oxidizing reagents, effect of surfactants, bleaches and enzymes can be detected on stain removal.

17. Bibliography

- **Books:**

1. "Lab Books." www.vernier.com, https://www.vernier.com/products/books/. Accessed Oct 26, 2017.

Journals:

Blanchard, A. A. "Synthesis of SODIUM PERBORATE - PrepChem.com." www.prepchem.com, Sept 19, 2015, http://www.prepchem.com/synthesis-of-sodium-perborate/. Accessed 20 Nov, 2018.

Oden, Matt. "The Science of Stain Removal | Remove Stains." remove-stain.com, http://remove-stain.com/science-stain-removal. Accessed Oct 3, 2018.

Robinson, Allan. "How to Use Sodium Perborate." sciencing.com, Apr 24, 2017, https://sciencing.com/use-sodium-perborate-5215007.html. Accessed Nov 5, 2017.

Websites :

"How to Make Luminol Glow: The Glowing Reaction." www.carolina.com, https://www.carolina.com/teacher-resources/Interactive/luminol-glowing-reaction/tr10786.tr. Accessed Nov 17, 2017

1. "Hydrogen Peroxide - the Evonik History Portal - the History of Evonik Industries." history.evonik.com, http://history.evonik.com/sites/geschichte/en/inventions/hydrogen-peroxide/pages/default.aspx. Accessed Jan 8, 2018.

"Laundrypedia Ingredients | Innovation - Ariel." www.ariel.in, https://www.ariel.in/en-in/about-ariel/innovation/laundrypedia-ingredients. Accessed Oct 8, 2017.

"Luminol - Dictionary Definition of Luminol | Encyclopedia.com: FREE Online Dictionary." www.encyclopedia.com, Feb 25, 2018, https://www.encyclopedia.com/science/encyclopedias-almanacs-transcripts-and-maps/luminol. Accessed Jan 11, 2018.

"meritnation.com." www.meritnation.com, https://www.meritnation.com/ask-answer/question/why-h2o2-is-better-oxidizing-agent-than-water-is-there-anyth/hydrogen/6828279. Accessed Nov 21, 2017.

"Organic Chemistry Portal." www.organic-chemistry.org, Feb 26, 2015, http://www.organic-chemistry.org/. Accessed Oct 10, 2017.

2. "PubChem Search." pubchem.ncbi.nlm.nih.gov, https://pubchem.ncbi.nlm.nih.gov/search/. Accessed Nov 3, 2017.

"https://www.sigmaaldrich.com/content/dam/sigma-Aldrich/docs/Sigma/Product_Information_Sheet/a8511pis.pdf. Accessed Nov 17, 2017.

- **Images:**

1. Principle of Spectrophotometer and Its Applications| Chemistry| Byjus." byjus.com, Jan 20, 2017, https://byjus.com/chemistry/spectrophotometer-principle/. Accessed Jan 11, 2018.

18. Appendix

Trial graphs for the compounds

1) Trial 2- 0.1g sodium per borate

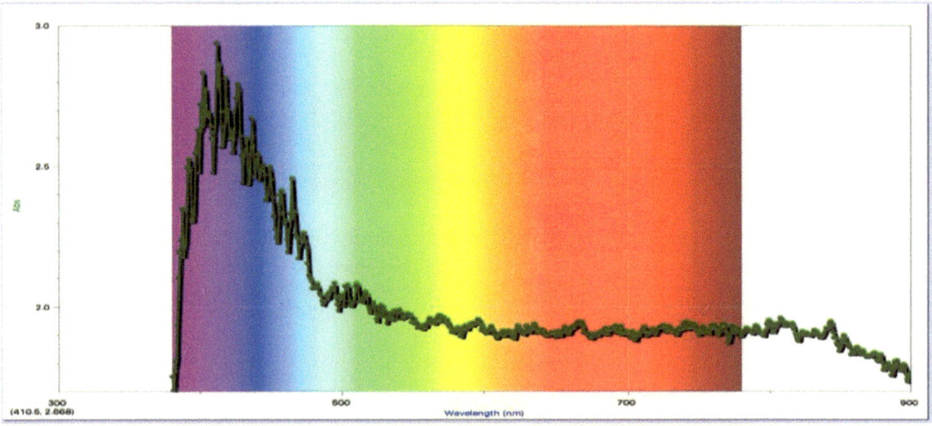

2) Trial 3- 0.1g Sodium per borate

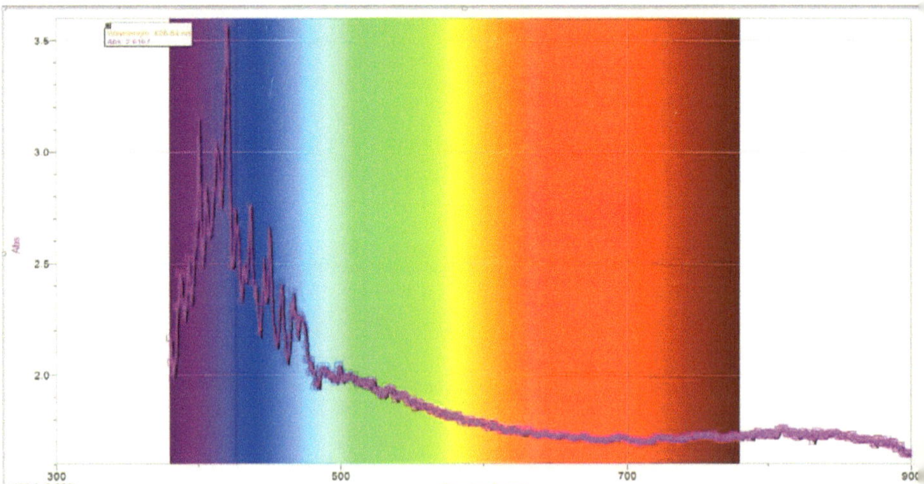

3) Trial 4 - 0.1g sodium per borate

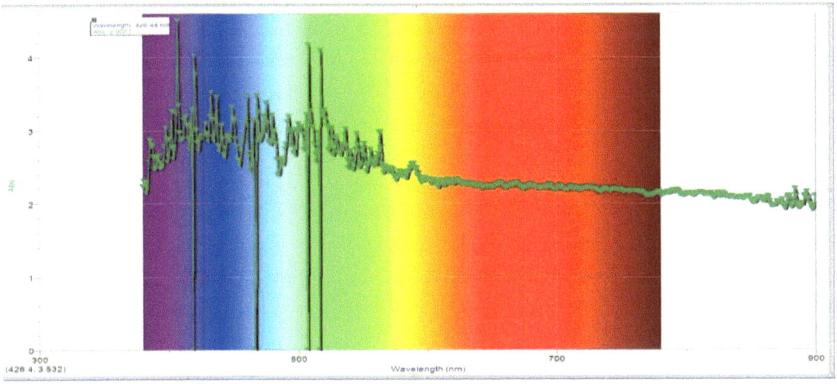

4) Trial 5- 0.1g of sodium per borate

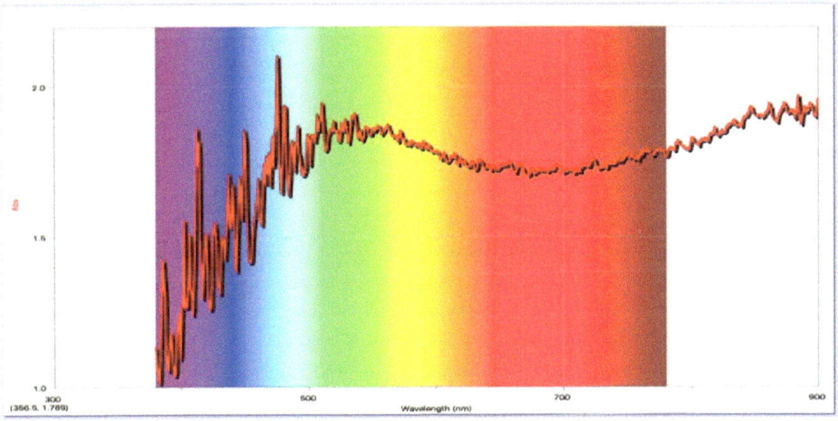

5) Trial 2- 0.1g of hydrogen per oxide

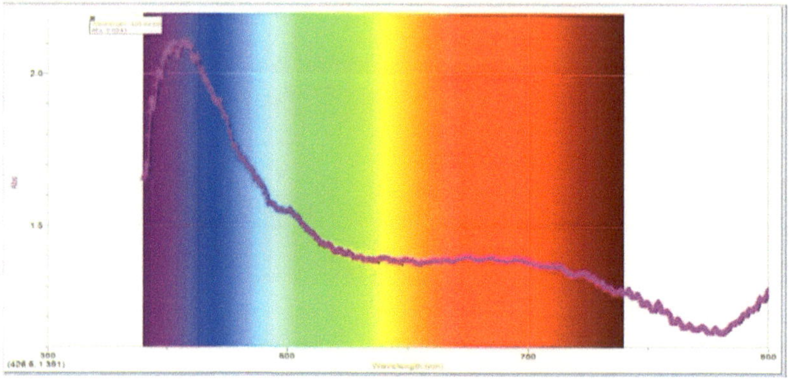

6) Trial 3- 0.1g of hydrogen per oxide

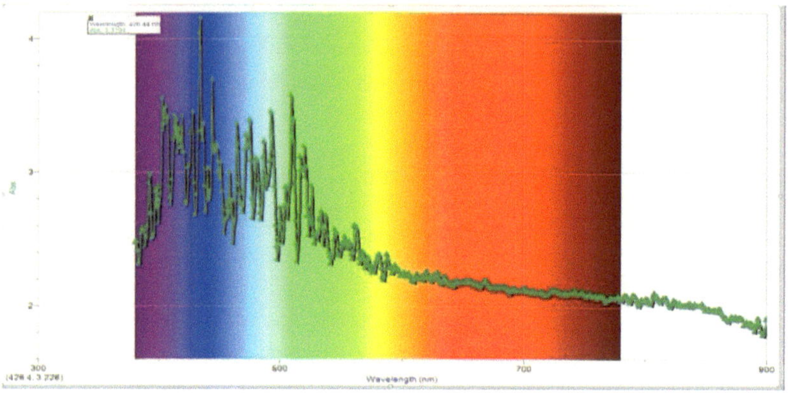

7) Trial 4- 0.1g of hydrogen per oxide

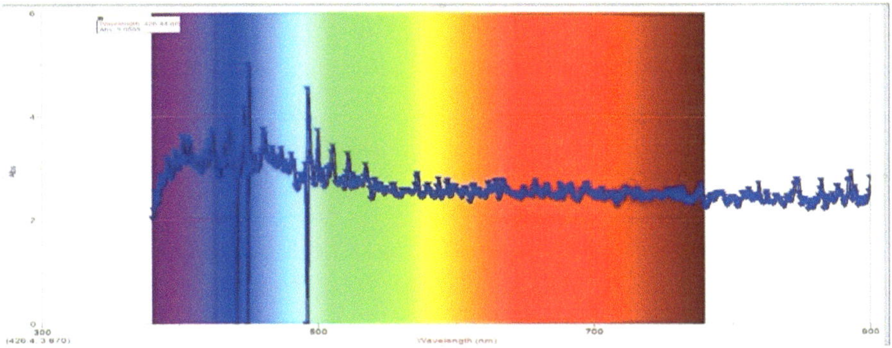

8) Trial 5- 0.1g of hydrogen per oxide

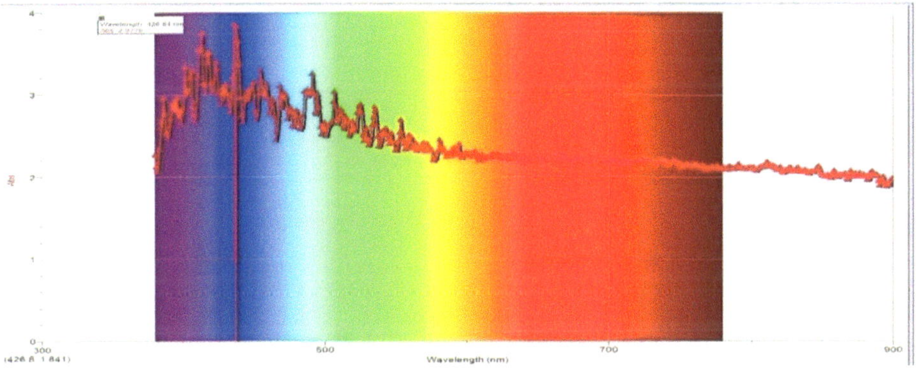

9) Trial 2- 0.1g of sodium per carbonate

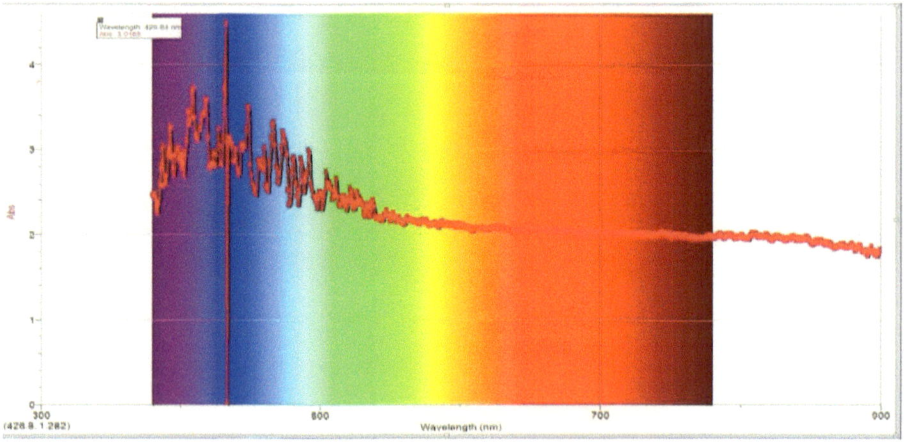

10) Trial 3- 0.1g of sodium per carbonate

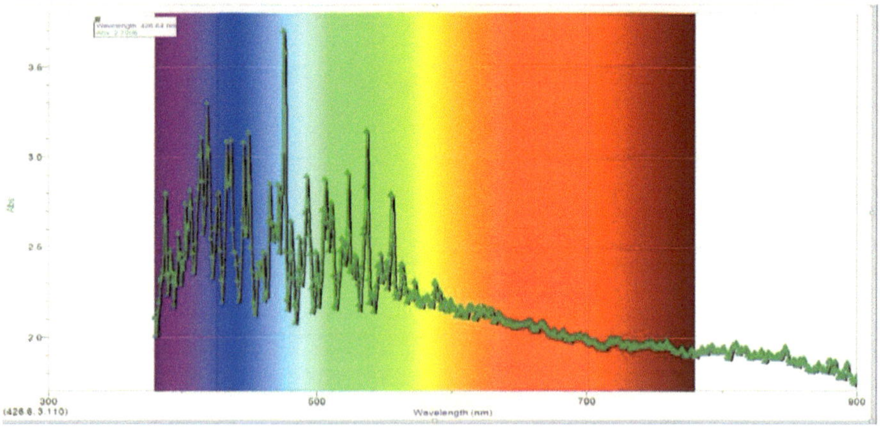

11) Trial 4- 0.1g of sodium per carbonate

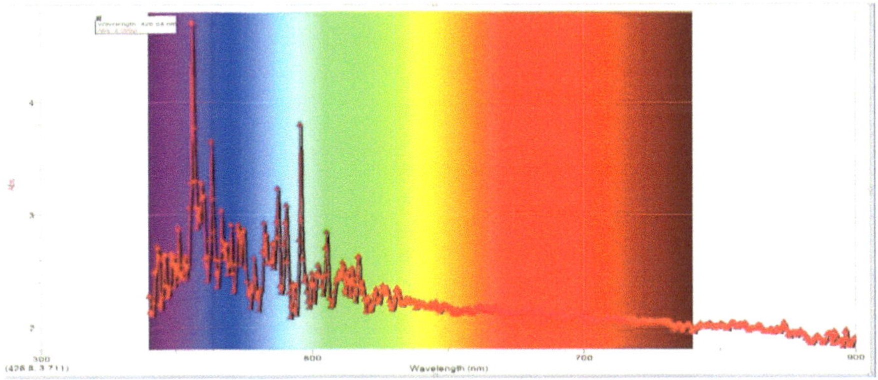

12) Trial 5- 0.1g of sodium per carbonate

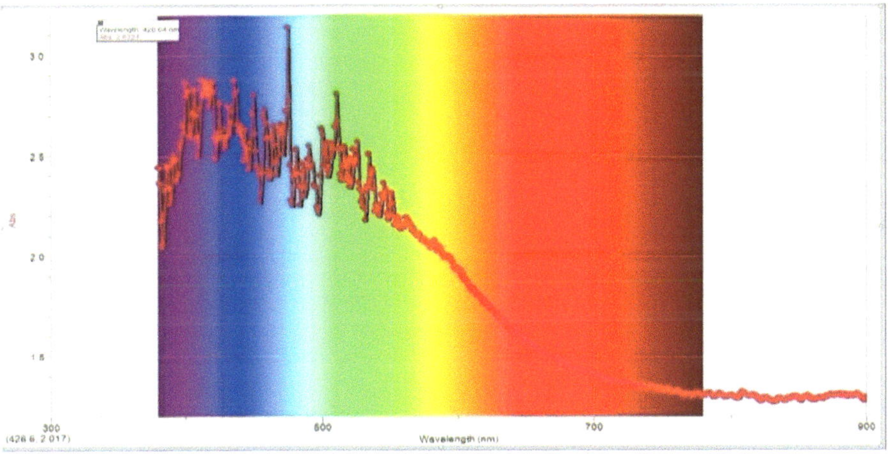

Graph 13 Absorbance shown by Luminol and Iron stock solution

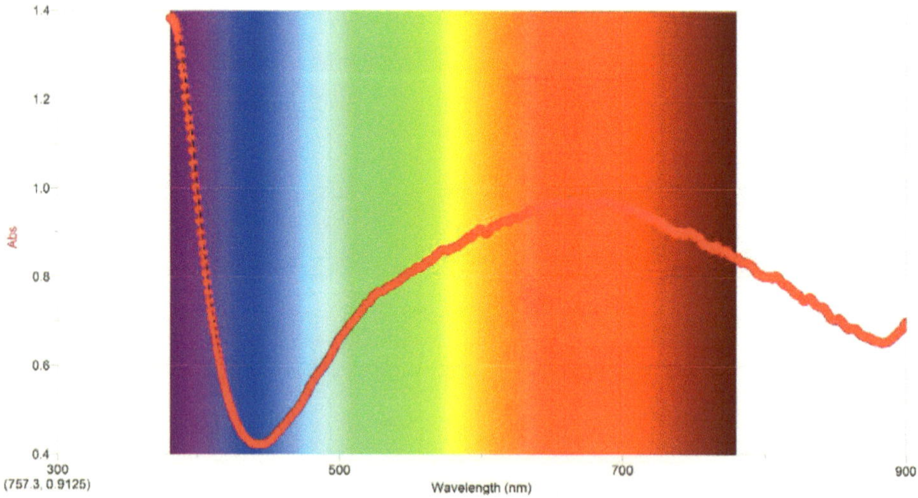

Absorbance data shown in tabular form

Wavelength (nm)	Absorbance shown (arb.unit)
420.3	0.429
421.2	0.428
422.1	0.427
423.1	0.426
424	0.425
424.9	0.425
425.8	0.424
426.7	0.424
427.6	0.423
428.5	0.422
429.4	0.422
430.3	0.423
431.2	0.423
432.1	0.425
433.1	0.426
434	0.429
434.9	0.431

Lowest absorbance shown

Experimentation performed on cloth:

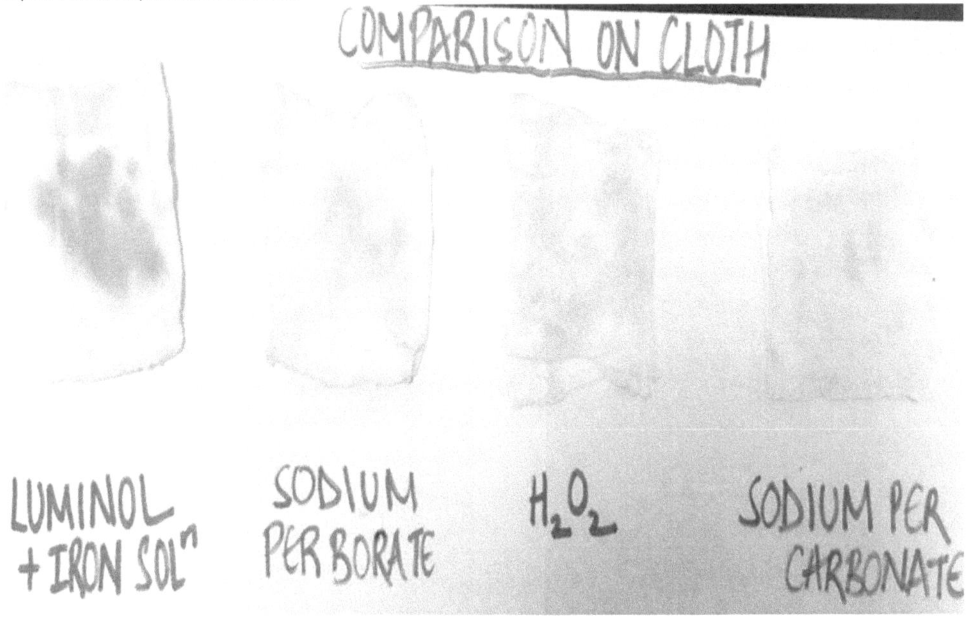